AMAZING
SCIENCE
TRICKS

AMAZING SCIENCE TRICKS

For **Kids and Parents**

Michio Goto

Translated by Tom Gally

KODANSHA INTERNATIONAL
Tokyo • New York • London

Distributed in the United States by Kodansha America, Inc., 575 Lexington Avenue, New York, N.Y. 10022, and in the United Kingdom and continental Europe by Kodansha Europe Ltd., 95 Aldwych, London WC2B 4JF.
Published by Kodansha International Ltd., 17-14, Otowa 1-chome, Bunkyo-ku, Tokyo 112-8652, and Kodansha America, Inc. Copyright © 1999 by Kodansha International Ltd. All rights reserved. Printed in Japan.

ISBN 4-7700-2492-4
First edition 1999
99 00 01 02   10 9 8 7 6 5 4 3 2 1

CONTENTS

III. AMAZING SCIENCE TRICKS FOR THE PLAY POOL ·········· 61

COLUMN 1 WHAT WERE THE GREAT SCIENTISTS LIKE WHEN THEY WERE LITTLE?
Richard Feynman 68

IV. AMAZING SCIENCE TRICKS AT THE DINING TABLE ········ 71

WHAT WERE THE GREAT SCIENTISTS LIKE WHEN THEY WERE LITTLE?
Albert Einstein   124

WHAT WERE THE GREAT SCIENTISTS LIKE WHEN THEY WERE LITTLE?
Thomas Edison   134

VIII. AMAZING SCIENCE TRICKS FOR OUTDOORS 169

IX. ADVANCED SCIENCE TRICKS 177

X. AMAZING SCIENCE TRICKS FOR FAMILY CONTESTS 185

PREFACE

Albert Einstein, the father of relativity, and the great inventor Thomas Edison started to learn about science when they were young children.

When Albert was given a compass by his father, the young boy was fascinated at how the needle always pointed north. Years later, the great physicist would describe how impressed he was by observing a force that he could neither see nor touch.

It was Edison's mother who gave him a book of science experiments for the home. By doing all the experiments himself, the future inventor learned how much fun science can be.

The purpose of this book is to help you show children how interesting—and fantastic—science really is. Despite the title, the 76 scientific tricks described in this book involve no deceit or sleight of hand. Each trick is straightforward and involves genuine scientific principles. Nevertheless, you can still surprise children with the incredible phenomena shown by the tricks. As with Einstein, the tricks will allow kids to experience forces that they cannot see or touch.

If you still remember your science lessons from your school days, you'll probably know immediately when a trick relies on air pressure or gravity or static electricity. But kids often know little about these forces, and even if they have heard of them, they will probably not realize how the forces affect everyday objects.

Children are sure to watch mesmerized when you do these tricks, and they'll want to know how they work. But please don't give an explanation right away. In my experience, it's better to first think about them together with the kids.

In Japan, many children approach science as just another subject for which they have to memorize a lot of facts in order to pass examinations. Recently, though, there have been signs of change. This trend started a few years ago, when teachers all over the country began to organize science fairs for youths. In just one location—the Science and Technology Museum in Tokyo—the science fair attracted as many as ten thousand families a day. Their numbers and enthusiasm show that many people are dissatisfied with the rote-learning approach to science education.

Another encouraging development has been the establishment, in 1998, of Science Ranger, a program sponsored by the Japan Science and Technology Corporation. Through this program, grass-roots science education is gaining ground throughout the country.

Some of the 76 scientific tricks in this book are my own invention, while others have previously appeared elsewhere. However, I have done all the tricks myself many times in order to simplify them so they can be easily performed by anyone.

In this way I have narrowed down the original one hundred tricks to their present number by excluding ones that were either too dangerous or required special equipment or complicated techniques. The resultant 76 tricks are the ones that I'm sure all adults and children will enjoy.

It is my sincere hope that the scientific tricks in this book will increase communication among people of all ages as well as contribute to the progress of society.

Michio Goto

A WARNING TO READERS

The tricks in this book are meant to be performed by adults. As some of them use materials that might be dangerous for children, please do not allow children to do these tricks by themselves.

Keep children in a safe place when they are watching the tricks. Be especially careful when using glass, sharp objects, fire, hot water, and other potentially dangerous things.

Also keep in mind that some of the tricks involve practice and may not work the first time, so be prepared.

I

AMAZING SCIENCE TRICKS AT
RESTAURANTS

Restaurants are great places to amaze kids with science tricks. There's often time to fill before your order arrives as well as after you've finished eating, and the items needed for the tricks are right there at the table. But don't try these tricks at a fancy restaurant. Save them for a friendly, family-oriented place.

Inseparable Napkins

They aren't glued or tied together, but you can't pull them apart.

Spread two cloth napkins on the table so that they overlap by about an inch.

Fold the overlapping section like a fan or an accordion. Pinch the folded section between your thumb and forefinger.

Ask a child to grab the two ends of the napkins and pull them apart. Even though you're holding the napkins with just your thumb and forefinger, the child will not be able to separate them.

How Does It Work?

Folding the overlapping sections of the napkins increases the degree of contact between the napkins, so that the pressure of your thumb and forefinger is enough to stop them from slipping apart.

What Else Can You Do? This trick can also be performed with handkerchiefs, washcloths, paper napkins, or just about anything made of cloth or paper.

Balancing a Quarter on a Dollar Bill

Place a quarter on the edge of any dollar bill and it won't fall off!

Fold a crisp dollar bill in half and place it upright on the table with the two halves at right angles. Carefully place a quarter on top of the fold.

Slowly pull the halves of the dollar bill straight.

The coin will move slightly as you pull the bill, but it won't fall off even when the bill becomes perfectly straight.

How Does It Work?

As you pull the dollar bill straight, the coin will move a little. The friction between the bill and the coin keeps the coin balanced. When the bill is completely straight, the coin's center of gravity will be directly over the edge of the bill.

To make sure this trick works well, use a crisp, new bill and pull the bill straight as slowly and carefully as possible.

A Mist-Making Device

Blow through a long straw held at right angles to a short straw, and watch the water in the short straw rise and turn into spray.

Order a soft drink to get a straw. Cut the straw into two sections, with the longer piece about twice as long as the shorter one.

Hold the shorter section upright in a glass of water. Hold the longer piece at right angles to the shorter one.

When you blow hard through the longer length of straw, a fine mist will spray out. (Be sure you do this trick with water, not a soft drink.)

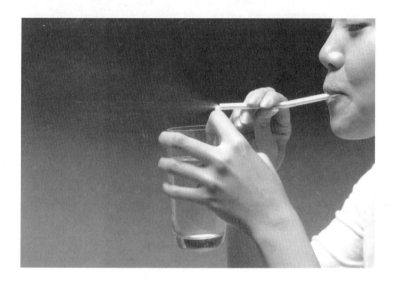

How Does It Work?

This is an example of the Bernoulli effect, when the air pressure drops as air moves quickly.

When you blow through the longer piece of straw, air flows quickly past the top of the shorter piece, making the air pressure inside it fall. The surrounding air presses down on the surface of the water in the glass, forcing water up through the short piece. This water is then turned into a mist by the pressure of the air coming out of the longer piece of straw.

Making a Pencil Spin

Rub a straw with a napkin to create a static electric charge of several thousand volts and use it to make a pencil spin around.

Balance a pencil on top of something with a smooth, round top, such as a sugar or ketchup container. If you're in a Japanese restaurant, you can use a wooden chopstick instead of a pencil.

Rub the straw briskly up and down five or six times with a dry paper napkin or tissue.

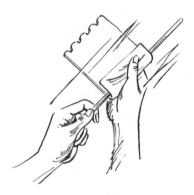

Hold the straw near one end of the pencil. The straw will pull the pencil toward it. By moving the straw in a circle, you can make the pencil spin around and around.

How Does It Work?

When you rub the plastic straw with paper, the straw receives a negative electric charge and the paper a positive charge. When you place the straw near one end of the pencil (which acts as an insulator), the pencil acquires a positive charge. Since positive and negative attract each other, the pencil is pulled toward the straw.

The electricity created by rubbing the straw with paper is called "static electricity." The amount of static electricity in one straw can be as much as several thousand volts.

A Balancing Trick with Forks

Although the forks and coin stick out beyond the rim of the glass, they don't fall off.

Insert a quarter between the middle prongs of a fork. Insert another fork to fix the coin firmly in place. This makes a balancing device with the coin as the fulcrum.

Place the coin on the rim of a glass, and adjust the angle of the forks so that the whole set-up is balanced.

How Does It Work?

The center of gravity of this balancing device is located along a vertical line passing through the fulcrum. When the forks tip in one direction or another, the center of gravity also shifts, but soon returns to directly below the fulcrum.

To make the trick appear even more amazing, drink some water from the glass while the forks are balanced on the opposite rim.

An Upside-Down Glass of Water

Surprise everyone by making the water in a glass "float" in the air.

Fill a glass with water. Cut a piece of paper slightly larger than the mouth of the glass and place it over the glass.

Press the paper down on the rim of the glass and slowly turn it upside down. Now take your hand off the paper. The water will stay in the glass without spilling. (Just in case, though, you should do this trick over a large bowl.)

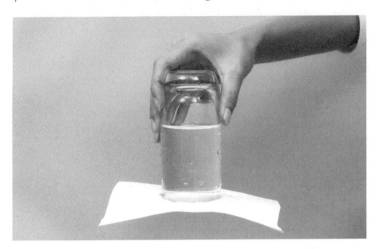

How Does It Work?

The water doesn't spill out because it is held in by air pressure and by the surface tension between the water and glass. The paper keeps the surface of the water flat so that air pressure is uniform.

A coaster will work even better than paper. If the glass is small enough, you can even use a credit card or phone card. It doesn't matter if the water contains ice, because ice is lighter than water. You can also do this trick with a glass that is half full. This makes the water appear to be "floating" in the air.

Two Full Glasses on Top of Each Other

Place one full glass of water upside down on top of another.

Fill two identical glasses with water.

Repeat the previous trick (page 24) with one of the glasses. When that glass is upside down, place it on top of the other glass.

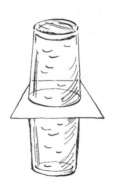

Holding both glasses firmly in place, slowly pull out the paper between them. The water in the top glass will stay inside the glass and not spill out. (Be sure to do this trick over an empty bowl, and clean up everything after you're finished.)

How Does It Work?

Even though there's a slight gap between the two glasses, this is sealed by the surface tension of the water. The pressure of the air outside also keeps the water from spilling.

An Upside-Down Bottle of Beer

If the restaurant serves bottled beer, order one. When you hold it upside down in a glass, the beer will stop pouring all by itself.

Hold an open bottle of beer upside down in an empty glass. The mouth of the bottle should be about halfway down the glass.

The people watching might get nervous as the foam rises in the glass, but the beer will automatically stop coming out.

How Does It Work?

This is another trick that relies on air pressure. The outside air pushes down against the surface of the liquid in the glass. This air pressure is offset by the pressure of the weight of the beer remaining in the bottle and the lower pressure of the air inside the bottle. When the balance is reached between the air pressure outside and the pressure inside, the beer stops coming out.

What Else Can You Do?

You can test this principle by using a simple device like the one shown here. Place two erasers side by side at the bottom of a shallow pan. Fill the pan with water to cover the tops of the erasers. Fill a clear plastic bottle with water, and place the bottle upside down on top of the erasers. The water won't spill out. When you scoop some water out of the pan, you'll see that the water level in the bottle drops by the same amount. This principle is used in bird feeders.

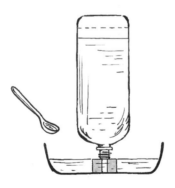

Catch the Twenty-Dollar Bill

Ask a child to try to catch the bill as you let go of it.

Hold out a twenty-dollar bill. Ask a child to make a V-sign with his or her index and middle fingers on either side of the bill. The fingers should be over the middle of the bill.

Say you'll give the child the money if he or she catches it when if falls. Then drop the bill.

If you catch it, it's yours.

Unless the child happens to guess exactly when to close his or her fingers, the bill will never be caught.

How Does It Work?

The average human reaction time—that is, the time it takes us to see something, assess it with our brain, and make the appropriate response, such as moving our fingers to catch a falling object—is 0.2 seconds. In the first 0.2 seconds after an object is dropped, it falls almost 8 inches. A twenty-dollar bill is just over 6 inches long, so by the time the child notices it falling, the top of the bill is already 5 inches below the child's fingers. This makes the bill impossible to catch.

Young kids and elderly people have especially slow reactions. That's also one reason they are more likely to get hit by cars when walking: even if they see a car coming, it takes them longer to react and get out of the way.

II

AMAZING SCIENCE TRICKS IN
^{THE} KITCHEN

The kitchen is a great place for scientific tricks because it doesn't matter if a little water gets spilled in the process. As long as you're responsible, you can even use the stove for your tricks. But be careful, and make sure that the kids don't get splashed with hot water.

A Plastic-Bag Porcupine

Stab pencils into a plastic bag filled with water without any water spilling out.

Fill a plastic bag with water and hold the top closed with one hand.

Puncture the bag with several newly sharpened pencils. The water won't leak out.

How Does It Work?

Most plastic bags are made of polyethylene. When polyethylene is heated, its molecules shrink, so when you stab a pencil through the polyethylene bag, the friction of the pencil against the plastic heats the polyethylene molecules and causes them to shrink around the pencil. That's one reason no water will leak out. Another reason is that the surface tension of the water keeps it from leaking out through the gaps between the pencils and the plastic.

A Straw That Bends Water

Place a straw near a thin stream of water from the faucet and watch the stream bend toward the straw.

Turn on the faucet so that the water comes out in a very thin stream.

Rub a plastic straw vigorously up and down several times with a paper tissue.

Hold the straw near the water and the stream will bend toward the straw.

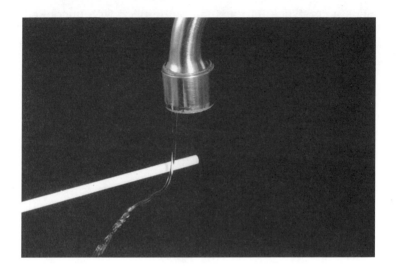

How Does It Work?

When you rub a straw with tissue paper, the straw acquires a high-voltage negative electric charge. Each molecule of water consists of two hydrogen atoms and one oxygen atom. The hydrogen atoms have a positive charge, while the oxygen atom is negative. As a result, the overall charge of each molecule is slightly more positive than negative.

When you put the straw near the stream of water, the positively charged water is attracted to the negatively charged straw. That's why the stream of water bends toward the straw.

Gripping Grains of Rice

Stick a chopstick into a small bottle filled with uncooked rice grains and lift the bottle up with the chopstick.

Fill a small bottle with uncooked rice grains. You can use any kind of small container as long as the mouth is narrower than the body.

Stick a single disposable chopstick halfway into the bottle, and press the rice down to pack it tightly. (You can buy cheap disposable chopsticks at many large supermarkets. They're also available at Asian restaurants and fast-food stands.)

When you pull up on the chopstick, the entire bottle will rise without the chopstick slipping out. (Just to be on the safe side, spread a towel under the bottle.)

How Does It Work?

Rice grains are smooth, but when they're packed into a bottle there's a surprising amount of friction between the grains and the chopstick. That friction is what prevents the chopstick from slipping out of the bottle when you pull up.

A Paper Cup
That Will Not Burn

Everyone knows that paper burns easily. But if you fill a paper cup with water, it won't catch fire even if you place it directly over a flame.

Insert two thin metal skewers through the upper portion of a paper cup. These will serve as handles.

Fill the cup halfway with water.

Turn the burner on the stove to medium, and hold the paper cup directly over the flame. The cup won't catch fire. (Make sure that the children don't put their faces too close to the flame.)

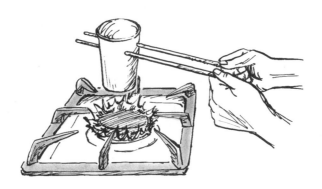

How Does It Work?

Paper burns only if its temperature is raised to several hundred degrees, which is much higher than the boiling point of water (212°F; 100°C). Even though water absorbs a lot of heat from the flame, the temperature of the water never gets hotter than its boiling point. As long as there's water in the paper cup, the cup can't get hot enough to burn.

A Tornado in a Bottle

An upturned bottle of water takes a long time to empty, but create a whirlpool effect by spinning the bottle, and the water drains out much more quickly.

Fill a two-quart plastic bottle with water.

Hold the bottle upside down above the sink. It might take as long as thirty seconds for all the water to drain out.

Now refill the bottle and turn it upside down again. This time, though, twist it around to create a tornado effect inside the bottle. Now the water will drain out much more quickly.

How Does It Work?

Look at the water as it drains out. When you make the tornado effect, there's an empty column at the center of the tornado. Air enters the bottle more easily through that column, so it allows the water to drain out faster.

What Else Can You Do?

The tornado may be easier to see if you cover the mouth of the bottle partially with tape. Leave a gap of about half an inch.

Fountains in a Bowl

Fill a metal bowl with water and rub the rim. Water will always splash up from the same four places.

Fill a metal bowl with water and place the bowl on a damp dishcloth to hold it steady.

First wash your hands with soap to remove any oils. With the palm of your hands, rub the rim of the bowl back and forth.

You'll see water splash up in four locations.

How Does It Work?

When you rub a metal bowl, it vibrates in a specific way. That vibration creates a resonance effect, with the vibration especially strong at four points. The water at those points vibrates especially strongly, so it actually splashes up. These four resonance points are characteristic of all metal bowls, so you'll always see the water splash up at the same locations.

What Else Can You Do?

You can create the same effect by rubbing the handles of a wok back and forth.

Bending Light Through Water

Although light is supposed to travel in a straight line, you can "bend" it by shining it through a stream of water.

Punch a hole in a clear plastic bottle 2 inches from the bottom. Put your finger over the hole, fill the bottle with water, and cap it to keep the water from leaking out.

Get a flashlight and darken the room. Cover part of the light with your fingers to make the beam narrower.

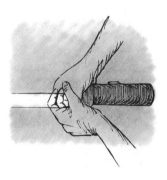

Take the cap off the bottle. The water will flow out in a parabolic stream. Shine the flashlight at the stream from the opposite side of the bottle. The light will bend with the stream of water and create a bright glow where the water hits the sink.

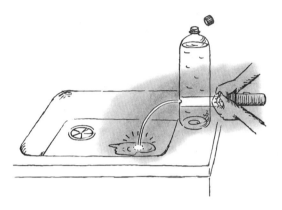

How Does It Work?

Why does some of the light bend with the water instead of continuing straight ahead? The answer is something called "total internal reflection." When the light in the stream strikes the boundary between the water and air, much of the light is reflected back into the stream. The light continues this internal reflection all along the parabola formed by the falling water. The same principle is used to transmit light signals through flexible optical fibers.

Making Colors in a Clear Plastic Bag

Shine a flashlight through a plastic bag filled with water and make the light turn red and blue.

For this trick you will need a long, narrow transparent plastic bag. Put a few drops of milk in the bag and fill it with water.

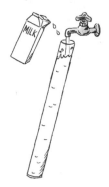

Tie the mouth of the bag and place it on a table. Darken the room and shine a flashlight lengthwise through the plastic bag. Make the beam of light narrow as explained in the previous trick (page 46).

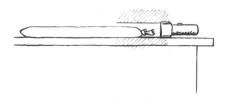

If you look closely, you'll see that the light in the bag is bluish on the side near the flashlight and reddish on the far side.

How Does It Work?

This trick works on the same principle that makes skies blue and sunsets red. The white light from the flashlight is a combination of red, blue, and many other colors. When the light passes through air or water, some of it bounces off suspended particles. Blue light has a short wavelength, so it reflects off the particles relatively easily. Red light, though, has a longer wavelength, so not as much is reflected. Thus most of the light reflected near the flashlight is blue, while the red light continues all the way through to the other end of the bag. The particles in this case are provided by the milk.

Stuck-Together Coffee Cups

Lift one cup up and the other cup will come up too.

Prepare two identical coffee cups. Cut a piece of newspaper 8 inches square, and fold it in four. Wet the newspaper with water.

Fill both cups about halfway with boiling water. Carefully pour all the hot water out of one cup, set it down, and cover it immediately with the wet newspaper. Then empty the second cup and quickly place it rim-down on the newspaper. Make sure that the rims of the two cups are aligned.

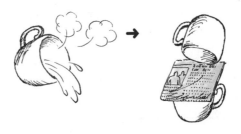

Wait about a minute, and then lift the cup on top. The other cup will rise with it. (Just in case, though, you should do this trick over a towel.)

How Does It Work?

When you empty the hot water out of a cup, the cup is full of water vapor instead of air. If you seal the cup promptly and let it cool, the vapor will condense into liquid again, so the pressure inside the cup will drop. The higher pressure of the surrounding air then holds the two cups together.

You can speed this trick up by pouring cold water over the cups to cool them more quickly. They will then stick together in just a few seconds.

An Imploding Can

Drop a heated aluminum can in cold water and watch it crumple.

Put a tablespoon of water into an empty aluminum soft drink can. Insert a skewer or a chopstick through the pull-tab to lift the can.

Hold the can above a flame for about 10 to 20 seconds. The water will soon boil and steam will start spewing out. (Be careful the hot water does not splash on anyone!)

Immerse the heated can top-down in a bowl of cold water. The can will implode with a loud crunch.

How Does It Work?

The principle is the same as with the coffee cups in the previous trick (page 50). When the water inside the can boils, the steam forces the air out of the can. When you cool the can, the steam condenses back into water. Since water takes up much less space than steam, the inside of the can becomes a near-vacuum. The pressure from the surrounding air then crushes the can.

What Else Can You Do?
You can use the same technique to crush plastic soft-drink bottles. Pour some hot water into a bottle and shake it well. Pour the water out and screw the cap back on the bottle. You can then crush the bottle just by putting it in cool water. You don't even need to use water—just leave the bottle standing, and it will gradually implode as it cools.

A Fountain in a Bottle

Stick two straws in a plastic bottle of water. Blow through one straw and water will spurt out of the other straw.

Fill a plastic soft-drink bottle about three-quarters full of water. Insert one straw into the bottle so that the end is in the water. Insert the other straw—a flexible straw works best—above the water.

Close up the mouth of the bottle with a wad of wet tissue paper to hold the straws in place and to create a seal.

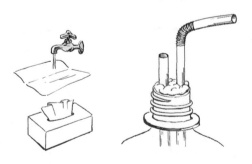

When you blow strongly through the straw that is above the water, water will shoot out of the other straw like a fountain.

How Does It Work?

When you blow into the straw, you increase the air pressure on the surface of the water. That pressure forces the water up and out of the other straw.

So what happens if you suck through the same straw? This time, the air pressure inside the bottle will decrease, so air will be drawn into the bottle through other straw. That air will create bubbles in the water.

A Soft-Boiled Egg, Backwards

A normal soft-boiled egg has a soft yolk and a hard white. Here's how to make one that has a hard yolk and a soft white.

Place an egg in a pot with enough water to cover the egg.

Using a cooking thermometer, heat the water to between 149°F and 154°F (65°C and 68°C).

Keep the water at that temperature for about 30 minutes. When you crack open the egg, you will find that the egg yolk is hard while the egg white is soft.

How Does It Work?

Eggs contain proteins that harden when they are heated, but the proteins that are in egg whites and egg yolks harden at different temperatures. Egg whites start to harden at about 158°F (70°C) and become firm only at temperatures above 176°F (80°C). But you can get egg yolks to harden at temperatures of about 149°F to 154°F (65°C to 68°C) if you're patient enough. So that's how this trick works: if you keep the temperature above 149°F and below 158°F, you can make a soft-boiled egg with a soft white and a hard yolk.

A Growing Egg

Soak an egg in vinegar for three days and make it grow one-and-a-half times in size.

Place an uncooked egg in a large glass, then fill the glass with enough vinegar to cover the egg completely.

Leave the egg like that for three days. During this time, bubbles will come out of the egg and the egg will slowly grow larger. (This is not the fastest trick in this book.)

After three days, the eggshell will have disappeared and the egg will be covered with a translucent film. It will be about 50 percent larger than before.

How Does It Work?

The eggshell dissolves by reacting with the acetic acid in the vinegar. Eggshell is made of calcium carbonate, which reacts with acetic acid to form carbon dioxide. That carbon dioxide is the source of the bubbles.

The reason the egg grew in size is related to something called "osmotic pressure." When two substances with different densities are next to each other and only separated by a membrane, osmotic pressure causes the densities to equalize. In this case, the membrane is the translucent film covering the egg after the shell has dissolved. Since the interior of the egg is denser than vinegar, moisture passes from the vinegar into the egg to equalize the density. That's why the egg gets bigger.

III

AMAZING SCIENCE TRICKS FOR
THE PLAY POOL

Some tricks work best when you have lots of water to play with and nobody minds getting wet. A backyard play pool makes a great "laboratory" for these tricks.

Perpetual Motion with a Hose

Isn't perpetual motion supposed to be impossible according to the laws of physics? But with just a rubber hose you can make water flow out of a pool continuously.

Get a length of hose four to five feet long.

Immerse one end of the hose in a pool of water. Place the other end in your mouth and suck until the water reaches your mouth. Remove the hose from your mouth and block the end with your thumb to stop the water flowing back into the pool.

Lower the end you're holding toward the ground so that it is below the level of the pool water. Take your thumb off the opening and the water will flow out of the pool in a steady stream.

How Does It Work?

When water pours out of one end of a hose, a vacuum is created inside the hose. To fill that vacuum, air pressure on the surface of the water in the pool pushes water up the hose. That water, in turn, forces more water out of the hose.

It may help to think of the water in the hose as an object. When you put the end of the hose below the surface of the pool, the water's center of gravity is outside the hose. This imbalance causes the water to flow out when you release your thumb.

You can do the same trick by taping several straws end to end. Make sure that the joins are sealed well. At least one of the straws must be the flexible type so that it can bend.

A Toothpick Torpedo

Put some shampoo on the tip of a toothpick and watch the toothpick move by itself.

Dab a little shampoo on the blunt end of a wooden toothpick.

Drop the toothpick in a pan of water. The toothpick will start moving in the direction of the sharp end.

How Does It Work?

Shampoo contains surface-active agents, which reduce the surface tension of liquids. As the shampoo on the end of the toothpick dissolves, it reduces the water's surface tension around it, thus releasing the water's hold on that end of the toothpick. The water around the other end of the toothpick still has surface tension, so it pulls the toothpick in that direction.

After you've done this experiment once, the entire surface of the water will be covered with a film of shampoo. If you want to repeat the trick, stir the water to break up the film.

Fantastic Soap Bubbles

Dip a cube made of pipe cleaners in soapy water and see the surprising shape that emerges.

Make a cube-shaped frame out of pipe cleaners. (These furry wires are available at toy stores or craft shops.) Attach a piece to one corner of the cube so you can suspend it.

Dissolve some soap or shampoo in warm water.

Immerse the wire cube completely in the soapy water. When you pull it out the shape of the "soap bubble" will be quite fantastic.

How Does It Work?

As long as the wire cube is the same shape, the soap bubble will be the same shape every time you do the trick. That's because the soapy film always takes the shape that requires the least amount of energy. This minimization of energy also means that the surface of the soapy film will be the smallest possible. If you change the shape of the cube or use other shapes, though, the soapy shape will come out differently.

This trick provides visual proof of the scientific principle that the most stable condition is the one that requires the least energy.

WHAT WERE THE GREAT SCIENTISTS LIKE WHEN THEY WERE LITTLE?

Richard Feynman's Father Taught Him the Importance of Experiments

The American physicist Richard Feynman made enormous contributions to the field of quantum electrodynamics. The story goes that when he was a child, he used to love playing with his toy wagon. He would put a ball in the wagon and pull it along with a piece of rope.

Then something happened that he could not understand. When he pulled the wagon forward, the ball did not move with the wagon, but rolled backward instead. And when he stopped the wagon, the ball would suddenly roll forward.

He asked his father why this should happen, but instead of explaining, his father suggested that they try an experiment together. Here's a similar experiment that you can try, too.

First, lay a track of drinking straws on the floor, parallel to one another and about half an inch apart. The track should be at least a yard long. Then cut an empty milk carton in half lengthwise and tie a piece of string to the top end. Place a marble inside the carton.

Now pull the carton over the straws. If you suddenly stop, what happens to the marble?

The young Richard Feynman did a similar experiment many times over, and asked his father to explain the principles involved. He and his father thought about them together and eventually came up with the principle of the relative motion between the marble and milk carton and the idea of inertia.

By approaching the problem in this way, Richard Feynman learned two concepts. First, an object might appear to be moving or might appear to be at rest; it all depends on the perspective of the viewer. That is what is meant by the relativity of motion. Second, a moving object tends to continue moving, while an object at rest stays still unless something causes it to move. That's called "inertia."

Thanks to his approach, Richard Feynman acquired a lifelong interest in all fields of science.

How about your kids? Will they wonder why the marble moves that way in the milk carton? Try it yourself. You don't even need a wagon or a milk carton on straws. You can use a toy truck or devise your own set-up. Show a child this phenomenon, and see if he or she shows any signs of genius.

AMAZING SCIENCE TRICKS AT
THE DINING TABLE

When dinner's over, it's time to clear away
the dishes and start the magic. There are
many fun and educational tricks you can
do with objects close at hand.

Suspending a Kettle in the Air

Common sense tells you that this can't be done, but you can suspend a large kettle in the air with only a pair of chopsticks.

Wind a rubber band tightly around the end of a pair of disposable chopsticks. Place the chopsticks beneath the handle of a kettle.

Take a single chopstick, break it in half, and jam one end against the rubber band and the other end against the knob on the kettle's lid. Adjust the angle of the handle so it is slightly tilted.

Hang the kettle by the chopsticks from the edge of a table. The kettle will seem to float in the air below the table. (If this trick doesn't work because the knob on the lid is too small, try jamming the chopstick against the base of the knob or the lip of the lid.)

How Does It Work?

Notice that part of the kettle is hanging under the table. That's because the kettle's center of gravity is directly below the support point where the chopsticks touch the table's edge.

If the kettle slips off the chopsticks, you can increase the friction by placing something like a small piece of sandpaper between the kettle and the chopsticks.

An Erupting Volcano

A spoonful of baking soda in a bottle of carbonated soft drink will make the bottle erupt like a volcano.

Open a bottle of carbonated soft drink.

Drop a teaspoonful of baking soda in the bottle.

In a moment or two, the bottle will "erupt" with foam. (It's a good idea to place a pan or dish under the bottle!)

How Does It Work?

Baking soda is made of sodium bicarbonate. When you put that sodium bicarbonate in a carbonated drink, it produces a large amount of carbon dioxide, which combines with the carbon dioxide already dissolved in the carbonated liquid to create the huge eruption.

What Else Can You Do?

Some candies are made with baking soda. Try dropping a few Smarties or something similar into a bottle of carbonated drink to see if the trick will work.

A Candle That Sucks Water

Watch the water in a saucer get sucked up into a glass!

Place a candle upright in the middle of a saucer.

Fill the saucer with water and light the candle.

Place a glass over the candle. When the flame goes out, the water in the saucer will get sucked into the glass.

How Does It Work?

When the candle is burning inside the glass, the heat makes the air expand, so some of the air escapes outside the glass. The candle goes out after it uses up all the oxygen, so the air inside the glass cools down. As it cools, the pressure inside the glass drops. Some of the carbon dioxide formed by the flame dissolves in the water as well, decreasing the pressure even more. The water outside the saucer is forced into the glass by the higher air pressure outside.

A Chattering Quarter

Make a quarter rattle just by holding a bottle.

Place an empty glass bottle in a refrigerator for an hour or so to cool it. If you're in a hurry, you can cool it more quickly by immersing it in a bucket of ice water.

Tape one side of a quarter to the top of the bottle so it won't fall. Make sure you dry the bottle first so the tape sticks.

When you hold the bottle in your hands, the quarter on top of the bottle will start rattling.

How Does It Work?

The warmth of your hands warms the air inside the bottle and makes it expand. This air pushes against the quarter until the quarter rises to allow some air to escape. But since the quarter is heavy, it soon falls back and clanks against the bottle. This happens again and again, causing the rattling sound.

Your hands should be as warm as possible, so you might dip them in warm water first. It also helps to wet the quarter to make a tighter seal between the coin and the bottle.

A Steel Ball That Can Pass Through a Dime

Drop a steel ball on a bottle covered with a dime and watch the ball drop into the bottle.

Find a steel ball—a ball bearing should do—and a glass bottle. Make sure that the steel ball is small enough to go through the mouth of the bottle. Place a dime over the mouth of the bottle.

Roll a sheet of letter paper into a cylinder and tape it to keep it from unrolling. Place it over the bottle and then drop the ball into it.

Even though the mouth of the bottle is covered with the dime, the steel ball will fall into the bottle.

How Does It Work?

When the steel ball hits the dime, they both bounce up. Because the dime is lighter than the ball, the dime bounces higher and so the ball is able to slip by it and drop into the bottle. The dime then falls back into place over the mouth of the bottle.

This trick works best when the paper cylinder is fairly long. It will not work as well with plastic bottles, because too much of the impact of the steel ball is absorbed by the plastic so the dime won't bounce as high.

A Can That Can "Walk"

Make an aluminum can roll toward a balloon.

Place an empty aluminum can on the floor.

Blow up a balloon and tie the end. Rub a tissue paper back and forth on the balloon.

When you put the balloon near the can, the can will start rolling toward the balloon.

How Does It Work?

When you rub the balloon with a tissue, the balloon gets a negative electric charge of several thousand volts. When you put the balloon near the can, electrostatic induction affects the molecules in the metal. The outside of the can gets a positive charge, so it is drawn toward the balloon and starts rolling in that direction.

What Else Can You Do?

Instead of a balloon, you can use things made of plastic. A plastic bag will work, and maybe even a straw. Just rub the plastic bag or straw with a tissue.

Making Sparks with Your Finger

Wind some plastic wrap around an aluminum can, take it off and watch the sparks fly as you put your finger near the can.

Place a flexible straw on top of an empty aluminum can and tape it down. The straw will serve as a handle so you don't touch the can directly.

Wrap a sheet of plastic wrap once around the can. Lift the can by the straw and hold it in the air. Now unwrap the plastic.

Put your finger near the can. A spark will fly between the can and your finger, and you'll get a small shock.

When you peel the plastic wrap off the can, you cause what is called "charge separation." This happens when two different substances are separated, causing both to acquire an electric charge. In this case, the aluminum can becomes charged with several thousand volts. This voltage is released when the can comes close to a person who is grounded, hence the sparks.

It's okay to let children do this trick. Although they'll feel a shock, the electric current in the spark is too small to do any harm.

Tasting Metal

If you hold a metal spoon and a piece of aluminum foil against your tongue so they do not touch, you will not taste anything. But bring the end of the spoon and the foil into contact, and your taste buds will be activated.

Ask a child to hold a metal spoon in one hand and a piece of aluminum foil in the other hand. Have the child place them against his or her tongue, taking care that the two don't touch each other. The child will notice no taste.

Now ask the child to touch the lower end of the spoon with the foil. The child will immediately notice a bitter taste.

How Does It Work?

When two different metals are put into an electrolyte (a liquid that can conduct an electric current) a battery is created. In this case, the electrolyte is the saliva on the tongue, and the spoon and aluminum foil are the two metals. When you bring the spoon into contact with the aluminum foil and hold both against your tongue, this "battery" creates a current that stimulates your taste buds. That's what causes the bitter taste.

A Magnetic Spoon

Magnetize a spoon by rubbing it with a magnet. Hit the spoon against a table and the magnetism vanishes.

Rub a metal spoon a few times in one direction with a magnet. You can use any kind of magnet—the magnetic note-holders for refrigerator doors will work fine.

Hold the spoon over some metal paper clips to check if the spoon attracts them.

Bang the spoon several times against the edge of a table. Now hold the spoon over the paper clips and nothing happens. The banging made the spoon lose its magnetism.

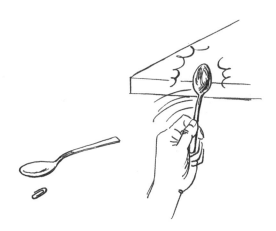

How Does It Work?

The metal in the spoon is made up of many tiny magnets. Normally, these little magnets are aligned in many different directions, so the spoon as a whole has no magnetism. When you rub the spoon with a magnet in one direction, you make the north and south poles of all the tiny magnets in the spoon face the same direction, so the whole spoon becomes a magnet too.

When you bang the magnetized spoon against a table or any other hard object, you scramble the directions of the little magnets. That's why the magnetism disappears.

Keeping Water Separate

Place a glass of colored water upside down on top of a glass of salt water. The two sorts of water will not mix.

This is basically the same as the trick on page 26. This time, add two tablespoons of salt to the water in one glass and stir well. Add a few drops of food coloring to the water in the other glass.

Cover the glass containing the colored water with a sheet of paper and turn it upside down. Place it on top of the glass containing salt water. (Be sure to do this trick over a dish.)

Gently pull the paper out from between the glasses. The colored water and the salt water will remain separate, with a clear boundary between them.

How Does It Work?

Salt water is heavier than colored water, so the two sorts of water stay separate as long as the boundary between them isn't disturbed. Try turning the two glasses over, though. The heavier salt water will now be on top, so it will flow down and mix with the colored water. Eventually the water in both glasses will be lightly colored.

What Else Can You Do?

Leave the two glasses overnight with the colored water on top and the salt water underneath. The two liquids will still stay almost completely separate. If you try the same trick with clear cold water on the bottom and colored warm water on top, the liquids will stay separate at first. However, as the warm water cools, its weight relative to its volume will become the same as that of the cold water, so the two will mix together.

Lifting a Film Canister Filled with Water

Place a business card over a plastic film canister full of water. When you lift the card gently, the canister rises with it.

Fill a plastic film canister to the brim with water. Cut four pieces of newspaper slightly larger than the canister. Wet the sheets and place them over the canister.

Press a business card firmly onto the newspaper. Squeeze the canister a little so some of the water overflows to seal it more tightly.

Use two fingers to lift the card. The film canister should come up, too. (Be sure to do this over a plate or a tray.)

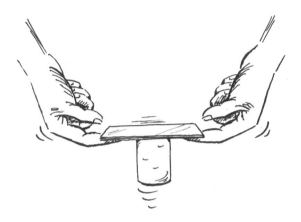

How Does It Work?

This trick relies on air pressure and on the surface tension between the water, film canister, and paper. Air pressure pushes the water and film canister up, and surface tension also pulls them up. These upward forces are stronger than the gravity pulling down on the water and canister, so the canister doesn't fall.

The pieces of newspaper help to strengthen the seal. With practice, you should be able to do this trick with just a business card.

Lifting a Jar with Your Palm

A heavy glass jar sticks to your open hand without falling.

Put a little hot water into an empty glass jar. Shake the water around a bit, and then pour the water out. (Be careful not to scald yourself!)

Place the palm of one hand firmly over the mouth of the jar.

As the jar cools, it becomes so tightly pressed onto your palm that no amount of shaking will dislodge it.

How Does It Work?

The water vapor emitted by the hot water forces the air out of the jar. As the jar cools, the vapor turns back into water and the pressure inside the jar drops. The palm of your hand makes a good seal, which keeps the outside air from entering the jar. The pressure of that outside air is what pushes the jar against your hand.

It feels as though your hand is being sucked up into the jar, but in fact it is the jar that is being pushed against your hand.

An Immovable Rubber Glove

No matter how hard you pull, you can't remove the rubber glove inside a milk carton.

Cut off the top of a milk carton with a paper cutter.

Put a rubber glove into the milk carton and fold the wrist part of the glove over the edges of the carton. Tape the edge of the glove to the outside of the carton with masking tape, making sure the inside of the carton is completely sealed.

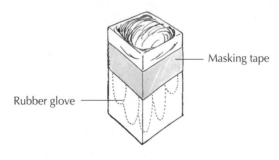

Masking tape

Rubber glove

Ask a child to pull the rubber glove out of the carton. No matter how hard he or she tries, the glove can't be pulled out.

How Does It Work?

The inside of the milk carton is completely sealed, so when you try to pull out the rubber glove, the interior of the carton becomes a near-vacuum. The pressure of the outside air on the glove is so strong that a child cannot pull it out.

Breaking a Chopstick with Air Pressure

Use a ruler to hit a chopstick held down by a sheet of newspaper. The chopstick breaks without the newspaper moving.

Place a disposable wooden chopstick (make sure it is completely dry) on a table so that one-third of it is jutting out over the edge. Spread a sheet of newspaper over the chopstick.

Pinch the paper firmly around the edges of the chopstick.

Use a heavy ruler or other object to give the chopstick a sharp hit just beyond the edge of the table. The chopstick will break, but the newspaper will not move.

How Does It Work?

This trick works because of the air pressure pushing down on the newspaper. Since there's no air between the paper and the chopstick, the chopstick is being pressed down by a strong force from above. This force is what keeps the chopstick from tipping up when you hit it quickly, so instead it breaks.

What happens if you try pushing the chopstick down slowly with your hand? The newspaper lifts easily. That's because air has time to get between the newspaper and the chopstick, so the air pressure from above is canceled out by the air under the newspaper.

Newspaper That Blossoms in Water

Put a folded piece of newspaper in water and watch it bloom.

Cut a square of newsprint 8 inches per side. Fold the four corners toward the center so that they meet in the middle.

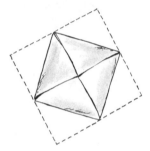

Cut out a flower shape from a piece of colored paper and insert it beneath the four corners. Gently place the newspaper in a bowl of water.

The corners of the paper will peel open to reveal the flower.

How Does It Work?

The fibers in the paper straighten when they become wet. That's why the folds will open. When the paper is completely soaked, it will lie flat on the water.

Melting a Styrofoam Tray

Squeeze the juice from a lemon peel onto a styrofoam tray and watch the tray melt.

Find a styrofoam tray of the sort used in supermarkets for packing meat and other foods.

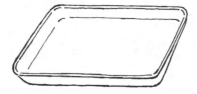

Squeeze the juice from a lemon peel onto the tray. The styrofoam will melt away. (Be sure to use the peel, not the pulp, of the lemon.)

How Does It Work?

Lemon peels contain a substance called "limonene" that melts polystyrene, the main component of styrofoam. This trick works not only with lemons but with other kinds of citrus fruit as well.

People have proposed using the same principle to dispose of styrofoam in garbage.

A Trapped Ice Cube

Keep an ice cube suspended between oil and water.

Fill a glass about halfway with water.

Fill the remaining half with oil. Any kind of vegetable oil or olive oil will do. The water and oil will remain separate, in two layers.

Drop an ice cube into the glass. It will float on the boundary where the oil and water meet.

How Does It Work?

This one is easy. Ice is lighter than water but heavier than oil.

An Egg That Floats

Eggs usually sink in water, but here's how you can suspend an egg in a glass of water.

Find a big glass and fill it halfway with water. Add salt to the water and stir. Keep adding salt until it doesn't dissolve any more.

Slowly pour in fresh water down the side of the glass so that it forms a layer on top of the salt water.

Gently drop an egg into the water. It will float motionless in the middle of the glass.

How Does It Work?

The water in the glass is divided into two layers, with the plain water on top and the water saturated with salt at the bottom. Since the salt water is denser than the egg, the egg floats above it. However, the egg is denser than the plain water, so it sinks below that. In this way, the egg stays suspended in the middle.

Here's how you can keep the plain water from mixing with the salt water. Take a milk carton and cut out a round piece the same diameter as the inside of the glass. Float the paper on top of the salt water, then add the plain water slowly from the top. The paper will keep the two types of water from mixing. Eventually, the paper will rise to the surface, and you can take it out.

A Boiled Egg Sucked into a Bottle

Place a boiled egg on top of a bottle and watch it get sucked inside.

Find a glass bottle that has a mouth slightly smaller in diameter than an egg. Pour some hot water into the bottle, shake it vigorously, and empty the water out. (Be careful not to scald yourself.)

Peel a boiled egg and place it on the mouth of the bottle. (Don't overcook the egg since it should be soft-boiled.)

Leave the egg on the bottle for a while and it will get sucked inside.

How Does It Work?

The vapor from the hot water drives the air out of the bottle. The egg then seals the top of the bottle, so air cannot get back in. As the water vapor cools, it turns back into water, so the pressure inside the bottle drops. The higher pressure of the outside air then pushes the egg into the bottle.

You can speed up the trick by cooling the bottle with cold water. It also helps if the water you use in the first step is as hot as possible.

A Fantastic Egg Drop

Pull the paper from under a cylindrical tube with an egg on it and the egg will drop straight into the water underneath.

Fill a glass three-quarters full of water. Cut a square from a milk carton slightly larger than the glass and place it over the glass.

Find a cardboard tube (the tube in a roll of toilet paper will do) and place it on top of the paper. Place a hard-boiled egg on top of the tube.

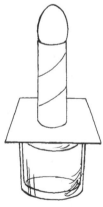

Quickly pull the paper away. The egg will fall into the water with a plop. (This trick doesn't always work, so be careful!)

How Does It Work?

When you pull away the paper, the lightweight cardboard tube flies off horizontally. Since the egg is heavier, it falls directly into the water below. This trick is even more exciting if you use a raw egg.

An Egg That Stands

Make an egg stand up. No gimmicks—it really works!

Place an egg on a table with the fatter end down. It doesn't matter if the egg is boiled or raw. Carefully adjust the angle of the egg until it stands upright.

With practice and patience, you can make the egg stand up. You don't need to smash one end of the egg or use any gimmicks. (It takes me about five minutes to make an egg stand up.)

How Does It Work?

If you look at the surface of an egg through a magnifying glass, you'll see that it is covered with many tiny bumps. If you can locate the egg's center of gravity directly above a triangle linking three bumps, the egg will stand up.

This trick takes some practice. You might want to try it first by spreading a paper tissue on the table and standing the egg on top of the tissue.

V

SHOW OFF YOUR ESP

Impress the kids with your ability to move and see things as if by magic. They'll think you have psychic powers—until you explain how the tricks work.

A Swinging Pendulum

Ask a child to choose one of three washers hanging from a thread and make that washer swing while the others remain motionless.

Take three metal washers and tie a thread to each. Cut the threads to three different lengths.

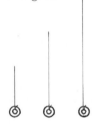

Tie the other ends of the threads to a stick. Ask a child to choose one of the washers.

Concentrate on the chosen washer and try to make only that one swing. With a little practice, you'll be able to do it.

How Does It Work?

The washers hanging from the threads are pendulums, and like all pendulums the frequency of their swings depends on how long the threads are. Longer pendulums have longer, slower swings, while shorter pendulums swing more quickly.

Suppose you want to make the washer on the shortest thread swing. All you have to do is move the stick back and forth at that pendulum's characteristic frequency. You can identify that frequency by moving the stick slightly until you find the speed that makes the smallest pendulum move. Once you have found it, push a little more in each direction to make that washer swing higher (this phenomenon is called "resonance"). Since the other pendulums have different frequencies, they will remain still.

Buildings and Earthquakes

Build three towers of different heights and make only one fall down.

With six aluminum drink cans, make three towers. Tape the two- and three-can towers together. (It doesn't matter if the cans are empty or full.)

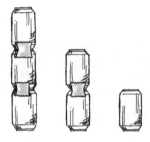

The three towers are make-believe buildings in an earthquake. Place all three on a piece of cardboard, and ask a child to choose one tower.

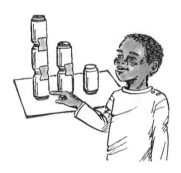

Focus your attention on the tower chosen by the child. Hold one side of the cardboard and shake it back and forth horizontally. With some practice, you'll be able to make only the tower chosen by the child fall down.

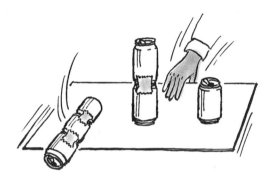

How Does It Work?

An earthquake knocks over a building if the frequency of the earthquake's tremors is the same as the frequency of the building's vibration. With these towers of cans, the frequency is higher for the shortest tower and lower for the higher towers. Thus, if you shake the cardboard quickly, its vibrations will resonate with those of the single can, so that can will vibrate more and eventually tip over. To knock over the taller towers, move the cardboard more slowly.

Snapping a Thread

Suspend a few metal washers between two lengths of thread and ask a child to choose the top or bottom thread. Then make that thread snap.

Tape together five metal washers. Tie two lengths of thread through the holes.

Hold the threads tight vertically so the washers are in the middle. Then ask a child to choose which thread should snap.

If the child chooses the top thread, pull the bottom thread down slowly. If the bottom thread is chosen, pull the bottom thread down quickly.

How Does It Work?

Which thread breaks seems to be pure chance, but in fact you can control it by how you apply force to the threads.

To break the top thread, pull down slowly on the bottom thread. Since both the pulling force from your hands and the weight of the washers are applied to the top thread, that thread will be the first to break.

To break the bottom thread, pull down on it sharply in order to apply as much force as possible. The inertia of the washers will keep that force from being conveyed immediately to the top thread. More force is applied to the bottom thread, so that one breaks first.

Reading Words Through an Envelope

No one can see into two envelopes—except you!

Ask a child to write a word of three large letters with a thick black felt pen on a piece of paper. The paper should be white on both sides.

Place the paper in a brown envelope, and insert that envelope into a white envelope. The writing on the paper should now be impossible to read.

Get a piece of dark construction paper, or tear out a page from a magazine that is printed on both sides. Roll up the paper into a 4-inch long tube. When you hold the tube against the envelope, you'll be able to read the writing inside.

How Does It Work?

Normally we cannot read the writing inside an envelope because of the light reflected off the envelope's white surface. The tube blocks off that reflected light, so you see only the light coming through the envelope. That's what enables you to read the writing on the paper.

WHAT WERE THE GREAT SCIENTISTS LIKE WHEN THEY WERE LITTLE?

Albert Einstein's Father Gave Him a Compass

The man who discovered the theory of relativity, Albert Einstein, was only five years old when he took his first step toward becoming a scientist.

One day when little Albert lay sick in bed, his father gave him a compass to amuse him. Albert was fascinated and excited at how the needle continued to point to north no matter which way the compass was turned. For the first time, he was aware of a force that, though impossible to see or touch, could affect things at a distance.

In his later years, Einstein would relate how that single incident had a profound effect that remained with him throughout his life. This was how he came to realize that deep principles lie hidden behind everyday phenomena.

It is easy to make your own compass with the magnetic buttons for refrigerator doors and a sewing needle. Rub the needle with the magnet several times in one direction to magnetize the needle. Now stick the needle through a small piece of styrofoam and place this in a glass of water. The floating needle will turn until it points north.

Since the needle is made of plated iron, it becomes a permanent magnet when magnetized and that is why it aligns in a north–south direction. One face of the magnetic button is the north pole and the other side is the south pole. If you rub the needle toward the sharp tip with the magnet's south pole, then the tip will become the needle's north pole and the end with the eye will be the south pole. That's because the magnet's south pole attracts the north poles of the tiny molecular magnets within the needle.

Demonstrate this with a child and see if he or she gets as excited as young Einstein did.

AMAZING SCIENCE TRICKS WITH
THE BODY

Some scientific tricks need no equipment at all—only our bodies. You may have done some of these tricks when you were a child. Now you can teach them to your kids.

The Iron Finger

Place one finger against the forehead of a seated child, and the child won't be able to stand up.

Ask a child to sit down on a chair. Press your index finger on the child's forehead.

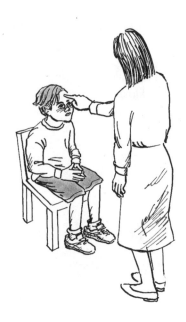

The child will not be able to stand up no matter how hard he tries.

How Does It Work?

When you stand up from a seated position, you first move your body forward to shift your center of gravity. Only then do you push against the floor with your legs to lift yourself up. If someone presses a finger against your forehead, then no matter how hard you try you won't be able to bend forward, so you won't be able to stand up.

The Shrinking Arm

Flex your arm and make it shrink an inch or so.

Ask a child to hold both arms stretched out in front. Show him or her that both arms are the same length.

Now ask the child to quickly flex one arm against his or her chest and straighten it again. Repeat this action about 30 times.

Compare the length of the arms again. The arm that was flexed will be an inch or more shorter.

How Does It Work?

The joints connecting our bones don't fit together tightly and are slightly loose. We can bend them because the muscles and tendons holding the joints are able to contract and expand. If you repeat the same motion many times, the muscles and tendons will contract, and it will take a little while for them to return to their normal length. That's why one arm is shorter after you exercise it. Don't worry, though. It will soon return to its normal length with no harm done.

Stretching Your Backbone

Even children who normally can't touch their toes can do so just by breathing out as they bend down.

Ask a child to place both feet together, keep both legs straight, and bend down.

If the child can already touch the floor, there's no point in doing this trick. Select a child who can only reach about 8 inches above the floor. Ask the child to exhale forcefully three times as he or she bends down.

Amazingly, each time the child breathes out, the hands will get closer to the floor. Soon the child will be able to touch it.

How Does It Work?

Breathing out forcefully loosens the muscles and tendons. This makes it easier to bend forward and touch the floor.

Thomas Edison's Mother Gave Him a Book of Science Experiments

When the great inventor Thomas Edison was a child, his mother gave him a book of science experiments that could be done at home, illustrated with easy-to-follow diagrams. The young Edison completed all the experiments in the book, and in the process learned how much fun science can be.

His curiosity and enthusiasm for experiments remained with him even when he became an adult. When he was looking for the right filament for his light bulb, Edison conducted over two thousand experiments. After many failures, he finally developed a bulb that would stay lit for a thousand hours.

The filament that Edison used for his light bulb was a piece of burned bamboo from the mountain called Otokoyama, above the Iwashimizu Hachiman Shrine in Kyoto Prefecture in Japan. The shrine is located in the city of Yawata, and in front of the station there is a road called Edison Avenue, where a bust of the famous inventor stands.

If you can obtain a piece of bamboo, here is how you can conduct the same experiment. First, place a long, thin strip of bamboo in a small sheet of aluminum foil and fold the ends over. Using tweezers, hold the foil over a flame for several minutes. Hot gas and flames will shoot out of the gaps in the foil. Allow the bamboo in the foil to cool for several minutes. When you remove it from the foil, it will have become a thin strip of charcoal.

Charcoal comes in two types—a soft, black type and a hard, white type. The soft type does not conduct electricity well, but the hard white type is a good conductor. If you connect four "C" batteries and wire them to the strip of burned bamboo, the white parts will light up just like the filament in a light bulb.

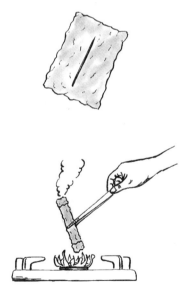

AMAZING SCIENCE TRICKS IN
THE LIVING ROOM

Instead of playing cards or video games, how about getting everybody together to do some scientific tricks?

A Musical Straw

Make a simple pipe with a drinking straw. As the straw gets shorter, the notes get higher.

Place one end of a drinking straw in your mouth, bite down on it, and blow. A note will sound, so now you have a musical straw.

Cut off small sections from the end of the straw while you are blowing. The notes will get higher with each cut.

How Does It Work?

The sound is caused by resonance inside the straw. The pitch of the sound is related to the length of the straw. A longer straw resonates at a lower pitch, while a shorter straw resonates at a higher pitch. If you cut the straw while you're blowing, then the pitch will rise.

What Else Can You Do?

Line up several soft drink bottles of the same size and fill them with different amounts of water. When you blow across the mouth of the bottles to make a sound, each bottle will have a different pitch. The principle is the same as for the musical straw. The pitch is determined by the resonance of the air above the water. The larger the volume of air, the lower the note.

A Hanging Belt

Even though the belt and pen cap extend beyond your finger, they won't fall off.

Slip a thin belt through the clip of a pen cap.

Place the pen cap carefully on the tip of your little finger. The cap and belt will stay suspended without slipping off.

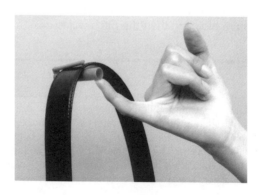

How Does It Work?

When the belt and pen cap are balanced, the center of gravity is located directly below the point of support. This keeps it from falling.

A Skewered Balloon

We all know that if we stick a sharp object into a balloon, the balloon will burst. Here's how you can pierce a balloon without that happening.

Find a thin wooden skewer that is normally used for barbecuing. Sharpen one end to make a fine point.

Inflate a balloon and tie the end. Slowly stick the skewer through the dark area opposite the tied end of the balloon.

The balloon won't pop. In fact, if the skewer is long enough, you can run it all the way through the knotted part at the other end. (The trick may not always work, so don't be surprised if the balloon pops.)

How Does It Work?

You've probably noticed that inflated balloons have a darker section around the bottom where the rubber is less taut. This part and the area around the tied end can be penetrated with a sharp object without the balloon popping. In fact, the rubber tightens around the sharp object to keep air from escaping.

A Floating Ping-pong Ball

A ping-pong ball placed on the cold air stream of a hair drier will bob up and down but will not fall.

Set a hair drier to "cold air" and switch it on. Point the drier upward, and carefully place a ping-pong ball in the path of the cold air.

The ball will bob vertically and horizontally, but it will stay within a fixed area without falling.

If you tip the hair drier at an angle, the ball will move in the direction of the air stream, but it will stop bobbing.

How Does It Work?

The air blowing at the ping-pong ball from underneath lifts it up, and the weight of the ball makes it bob up and down. The sideways motion is caused by differences in the speed of the air passing by the sides of the ball. When air is moving more quickly, the pressure drops—this is the Bernoulli effect. Therefore air pressure pushes the ball toward the side with the faster air stream.

The ball stays up when you tip the hair drier at an angle because the air stream now provides lift that compensates for the downward force of gravity. The streams of air passing to the right and left of the ball are the same speed, so the ball stops moving back and forth.

How to Balance Anything

Locate the center of gravity of an irregularly shaped object and balance it in three seconds!

Roll up newspaper or construction paper to make several narrow cones. Tape them together to form one long cone. The longer the cone, the more impressive the trick.

Support the cone on your index fingers placed under each end of the cone. Gradually move your fingers toward the center of the tube.

Slide your fingers under the cone bit by bit until they meet exactly at the cone's center of gravity. The cone will stay balanced at that point.

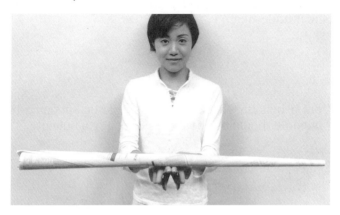

How Does It Work?

The center of gravity is near the thick end of the cone, so the finger holding up the narrow end is subjected to less friction from the paper. As you slowly move both fingers toward the center, the finger on the low-friction end will move first. As it moves, though, the weight on it increases until it stops moving, so the other finger, which now is under less pressure from the paper, starts to move. This alternation between the two fingers continues until they meet at the center of gravity.

What Else Can You Do?

If you use your arms instead of your fingers, you can balance just about any long object. If you're really strong, you can even find the center of gravity of a child lying straight across your extended arms.

A Dancing Butterfly

Make a plastic butterfly dance in midair by moving a plastic sheet underneath it.

Cut out a butterfly shape from a thin plastic shopping bag. The butterfly should be about the size of a real butterfly.

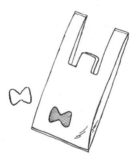

Find a sheet of firm plastic or a plastic board. Rub the sheet and the butterfly with tissue paper.

Throw the butterfly in the air and hold the plastic sheet beneath it. The butterfly will flutter in the air above the sheet. (Move the sheet and do not let the butterfly touch it or else it will stick to the sheet.)

How Does It Work?

When you rub the plastic butterfly and plastic sheet with tissue, both get a static electric charge of several thousand volts. The charge is negative for both, so they repel each other. The butterfly is light enough to be held up by this charge. What makes the butterfly flutter is the movement of the plastic sheet.

What Else Can You Do?

You can also rub a straw with tissue paper and use that to make the butterfly flutter in the air.

Lift a Child Up with Your Breath

Blow into a plastic bag and you can lift a child sitting on a board.

Place a large plastic garbage bag on the floor. Put a sturdy square board on top of the bag.

Place some books or other form of support under the board on the side away from the mouth of the plastic bag. Ask a child to sit on the board, then start blowing into the plastic bag.

You'll be able to lift the child and the board easily with your breath alone.

How Does It Work?

The weight of the child and the board is distributed across the entire area in contact with the plastic bag, so the weight on the bag at any single point is not very large. When you blow into the bag, you only have to apply enough pressure to lift the weight at one point of the bag. That's why it's so easy to lift the child.

If the child were standing directly on the plastic bag, it would be impossible to lift him or her with your breath.

Zero Gravity in a Paper Cup

Drop a paper cup, and the two erasers hanging outside it will be pulled in.

Prepare a paper cup and two small erasers about an inch long. Tape a thick rubber band to each eraser with masking tape. The rubber bands, when slack, should be shorter than the height of the cup.

Tape the other ends of the rubber bands to the bottom of the cup. Pull the erasers so they hang outside the cup.

Hold the cup up with one hand, drop it, and catch it with your other hand. The erasers will be pulled into the cup as it is falling.

How Does It Work?

Before you drop the cup, the elastic pull of the rubber bands is offset by the weight of the erasers, so the erasers remain hanging outside the cup. When the cup is falling, the weight of the erasers drops to zero, so the rubber bands are able to pull the erasers into the cup.

Pinhole Glasses for Nearsightedness

A pinhole in a piece of paper enables nearsighted kids to read far-away signs.

Make a pinhole in a piece of paper. Ask a child who is nearsighted to look through the hole at a sign in the distance.

The child will be able to make out letters that he or she can't see normally.

How Does It Work?

Objects viewed through a small hole appear sharp regardless of their distance because the hole focuses the light. When light from an object (such as light originating at points A and B in the diagram) passes through a small hole, the diameter of the light beam is kept narrow, so the image remains sharp and in focus.

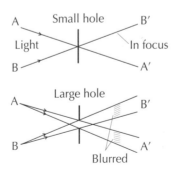

An On-again, Off-again Radio

When you put a radio inside a tunnel made of aluminum foil, the radio will stop working.

Use aluminum foil to make a tunnel large enough to cover a radio.

Switch on the radio and tune it to a station. With the radio playing, push it into the tunnel. The sound will automatically cut off.

Pull the radio out and it works again.

How Does It Work?

Radio waves cannot pass through metal or other materials that conduct electricity. The earth and concrete used in highway tunnels are also conductors, so they block off radio waves, too. That's why you usually lose radio reception when driving through tunnels.

An Immovable Business Card

Take a business card and fold it, and you won't be able to blow it off the table.

Fold a business card in the shape of an upside-down U.

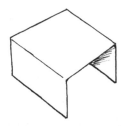

Blow as hard as you can through the opening under the card. The card will remain on the table without moving.

How Does It Work?

Because of the Bernoulli effect, the air pressure drops as the speed of air flow increases. When you blow through the opening under the card, you decrease the air pressure under the card, so there's more pressure from above holding the card against the table. That's why the card doesn't move.

If you're camping outdoors in a tent and the wind threatens to blow the tent down, just open the tent at both ends so that the wind can blow through it. This will stabilize the tent and keep it from blowing away.

Hot Balloon, Cold Balloon

Let a child feel how hot a balloon gets while you blow it up. Let the air out again and the balloon cools down dramatically.

Hold a balloon against the cheek of a child and blow it up. The child will be surprised at how hot the balloon gets.

With the balloon still held against the child's cheek, slowly let the air out. The child will be amazed at how cold the balloon suddenly gets.

How Does It Work?

As air is compressed, its temperature rises. You may have noticed what happens when you pump up a bicycle tire. As you pump in air, the metal nozzle on the tire gets warmer.

Blowing up a balloon also involves the compression of air. The energy that you supply by blowing is converted into heat, making the balloon warmer.

When you let the air out of a balloon, the thermal energy of the air is converted into motive energy, so the temperature drops.

A Revolving Thread

Make a thread turn around and around when you blow through the straw.

With a paper cutter, cut out a small square from the middle of a straw.

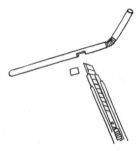

Insert a thread about three feet long into the opening and pull it out from the other end. Knot the ends of the thread and trim the ends.

Blow hard through the straw. The thread will begin to turn around and around through the hole.

How Does It Work?

When you blow through the straw, you create a rapid stream of air through it. This reduces the air pressure around the hole you made in the middle of the straw. The lower pressure draws the thin thread into the hole. As long as you keep blowing, the thread will keep turning.

Flying Paper Clips

Make two paper clips attached to a long strip of paper link up and fly through the air.

Cut out a long strip of scrap paper, 1 1/4 inches wide. Bend it into an S shape and clip it together with paper clips as shown.

Pull the ends of the paper, and the two paper clips will link up and shoot into the air.

How Does It Work?

If you pull the ends of the paper slowly, you'll see that the clips are gradually pulled closer until they become interlinked. When you pull the paper apart quickly, the linked clips shoot into the air.

 What Else Can You Do? As shown in the picture, you can do the same trick with three paper clips or even link two paper clips and a rubber band.

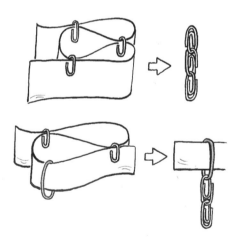

The Möbius Strip

Do you know how to turn one loop of paper into two interlinked loops? Just cut along the center of a Möbius strip.

Cut out a long strip of scrap paper. Give one end a half turn and tape the two ends together. Now you have a Möbius strip.

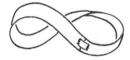

Cut the strip in half lengthwise to make one large loop.

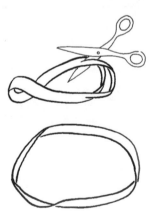

Now cut this loop in half lengthwise. You'll end up with two interlinked loops.

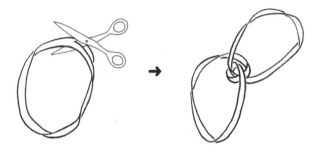

How Does It Work?

There's no real secret behind this trick. This is just what happens when you cut a Möbius strip.

The Möbius strip has many amazing properties. One is that it has only one side. That means that you can draw a line in one direction along the strip and return to your starting point without lifting your pencil off the paper.

What Else Can You Do? Try cutting the two loops in half again. You'll end up with a chain of four connected loops.

AMAZING SCIENCE TRICKS FOR
OUTDOORS

You don't have to stay indoors to do science tricks. Here are some neat tricks you can do out in the sunshine, too.

A Rainbow in the Mist

If you stand with your back to the sun and spray water into the air, you'll see a beautiful rainbow.

Stand with your back to the sun and spray water into the air with an atomizer.

You'll see a beautiful rainbow in the mist.

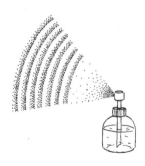

The mist you create with the atomizer consists of many tiny droplets of water. When sunlight strikes the droplets horizontally, the light enters each droplet and reflects back as shown in the diagram. Since light bends, or refracts, through water at different angles depending on the wavelength, the various colors in white light become separated from each other. That's what causes the rainbow.

Red light refracts at the highest angle to the ground, 42 degrees, so it appears at the top of the rainbow. Purple refracts at 40 degrees and appears at the bottom. (The angles in the diagram are exaggerated.)

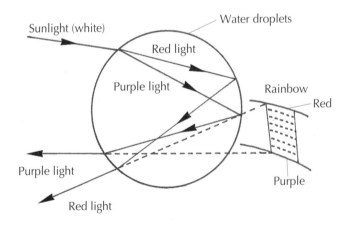

A Gauze Water Filter

Place one end of a piece of gauze in a glass of dirty water and the other end in an empty glass. Soon the empty glass will fill up with clean water.

Put some dirt into a glass of water.

Put an empty glass next to this glass. Place one end of a clean piece of gauze into the glass of dirty water and the other end into the empty glass.

Soon the empty glass will start to fill up with clean water.

How Does It Work?

If you watch closely, you can see the water start to rise through the gauze. This is caused by the capillary effect of the gauze's fibers. Only the water, not the dirt, is drawn through the fibers, so the water that flows into the adjacent glass is clean.

A Flying Trash Bag

Place an inflated black trash bag in the sun and watch it rise up.

Hold the mouth of a black trash bag in one hand. Use a hair drier to blow hot air into the bag.

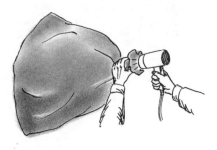

Seal the mouth of the bag with tape. Tie a long piece of string around the tape so you can hold it.

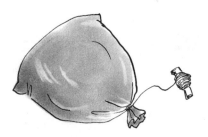

Take the bag out into the sun. The bag will rise slowly into the air. (It's best to do this trick in an open area on a windless day.)

How Does It Work?

Since the bag is black, it absorbs heat from the sun. That heat makes the air inside the bag expand and become lighter. When the bag and the air inside are lighter than the surrounding air, the bag starts to rise.

IX

ADVANCED SCIENCE
TRICKS

The tricks we've seen so far have been pretty simple but are still very impressive for children. Now let's try a few tricks that are more complicated and even more impressive.

A Cup That Empties

This trick is just right for people who don't want to drink too much. If there's too much liquor in the cup, most of it will spill out the bottom.

Make a hole in the bottom of a paper cup. Stick a flexible straw through the hole so that the bent end is inside the cup and turned down.

Do the trick with water instead of liquor. Pour water slowly into the cup. No water will spill out as long as part of the straw is still above the water.

As soon as the water covers the top of the straw, it will start pouring out.

How Does It Work?

When the U-shaped section of straw is completely immersed, the entire straw fills with water. The weight of the water makes it flow out of the cup. Once the water starts flowing, the air pressure from above continues to push it out until the water level drops below the straw's opening.

Flush toilets operate on the same principle.

What Else Can You Do?

You can make the trick even more surprising if you tape the straw to the inside wall of the cup and hide it with white paper.

A Smoking Cannon

By hitting the sides of a cardboard box, you can extinguish a candle ten feet away.

Use masking tape to seal a cardboard box. Cut a hole 4 inches in diameter in one end of the box.

Light several sticks of incense and hold the burning ends inside the box until the box is filled with smoke. Place a candle ten feet from the box directly in front of the hole. Light the candle.

Give a sharp slap on the sides of the box with your hands. A doughnut-shaped smoke ring will fly out of the hole and extinguish the candle.

How Does It Work?

When you hit the sides of the box, the volume of the box decreases. The air that is forced out takes the shape of a ring. This ring shoots ahead until it reaches the candle, extinguishing it. As the ring travels, the air rotates from the inside to the outside as shown in the diagram.

An Obedient Straw

Make the straw in a bottle of water rise and sink at your command.

Cut a 2-inch length of straw, bend it in a U shape, put a little water inside, and clip the ends together with a paper clip. Place the straw in a glass of water to test that it floats upside down near the surface.

Fill a plastic soft drink bottle with water. Place the straw upside down in the bottle and screw on the cap.

Shout "Sink!" as you squeeze the sides of the bottle, and the straw will sink. Shout "Come back up!" as you release your grip, and the straw will rise.

Sink!

How Does It Work?

When you squeeze the bottle, the pressure from your hands is conveyed through the water to the straw. The volume of the air in the straw therefore decreases, so the straw loses its buoyancy and starts to sink. When you let go, the air expands and the straw rises again.

This trick relies in part on Pascal's law, which states that a confined fluid transmits externally applied pressure equally in all directions. It also uses the Archimedian principle that an object immersed in a liquid undergoes an apparent loss in weight equal to the weight of the fluid it displaces.

X

AMAZING SCIENCE TRICKS FOR
FAMILY
CONTESTS

Here are two simple tricks that you have
to practice to do well. See who can mas-
ter them first, you or the kids.

A Free-Fall Pen

Make a pen balanced on top of a circular hoop fall straight into a bottle.

Place a circular hoop 6 inches in diameter on top of a glass bottle. An embroidery hoop should work fine.

Carefully place a pen upright on top of the hoop. (A short pencil or similarly shaped object will also work.)

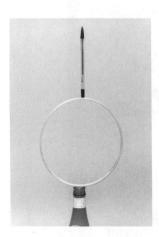

Grab the hoop and quickly pull it away. With practice, you can make the pen fall straight into the bottle.

How Does It Work?

This trick relies on the principle of inertia, according to which objects resist any attempt to put them in motion. If you pull the hoop away quickly enough, the pen's inertia will keep it vertical, thus allowing it to drop straight into the bottle.

The secret of this trick is to pull the hoop away as fast as possible. It's really fun when you succeed.

A Milk-Carton Boomerang

Make a flying device out of a milk carton, and train it to fly back to you.

Cut four strips out of a milk carton, each one 8 inches long by 1 1/4 inches wide. Glue two strips back to back to make two thick strips.

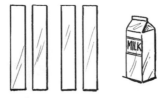

Place the two strips in the shape of a cross and tape them together. Bend the tips of the four spokes into ridges.

Hold the boomerang with the ridges facing you and throw it toward the ground. With practice, you can make it fly in a circular arc until it returns to where you're standing.

How Does It Work?

The ridges on the spokes increase the speed of the air flowing past the spokes' surfaces. Because of the Bernoulli effect, the pressure decreases on that side of the spokes, so the boomerang veers to the side where the ridges are. That greater pressure on one side of the boomerang is what makes it circle around to come back to you.

AFTERWORD

Of the many sources I used when writing this book, I would particularly like to mention *A Demonstration Handbook for Physics* by G. D. Freier and F. J. Anderson (published by the American Association of Physics Teachers). It first appeared nearly three decades ago, and it includes some eight hundred physics experiments gathered from all over the world, illustrated with detailed drawings.

In the introduction, the authors wrote, "A few of the demonstrations have been invented by the authors, but most of them have been invented by other teachers." Unfortunately, as most of the ideas are old, it is an impossible task to credit the inventors. It seems that in the area of teaching, the inventor of a teaching method tends to remain anonymous.

I also learned much from the newsletter "Galileo Study" and the group organized by Yoji Takikawa of the International Christian University High School in Japan. As a member of that group, I attend its monthly meetings with some fifty others. The unique experiments demonstrated at those meetings are published later in the group's newsletter.

Many enjoyable experiments contributed by teachers at schools and universities throughout Japan also appear in the guidebooks to the science fairs for youths that have been held nationwide since 1992. As chairman of the nationwide planning committee of the fair for the first four years, I was able to meet many active young teachers and take part in the development of many experiments.

I would like to take the opportunity here to thank all these teachers again for their help and cooperation.

SCIENTIFIC CONCEPTS DEMONSTRATED IN THIS BOOK

I. AMAZING SCIENCE TRICKS AT RESTAURANTS

Inseparable Napkins—friction

Balancing a Quarter on a Dollar Bill—balance of forces, center of gravity

A Mist-Making Device—air flow

Making a Pencil Spin—static electricity

A Balancing Trick with Forks—balance of forces, center of gravity

An Upside-Down Glass of Water—air pressure, surface tension

Two Full Glasses on Top of Each Other—air pressure, surface tension

A Upside-Down Bottle of Beer—air pressure

Catch the Twenty-Dollar Bill—falling motion, reaction time

II. AMAZING SCIENCE TRICKS IN THE KITCHEN

The Plastic-Bag Porcupine—polymer materials

The Straw That Bends Water—static electricity

Gripping Grains of Rice—friction

A Paper Cup That Will Not Burn—heat capacity of water, burning temperature of paper

A Tornado in a Bottle—whirlpool effect

Fountains in a Metal Bowl—resonance

Bending Light Through Water—total internal reflection

Making Colors in a Plastic Bag—light reflection

Stuck-Together Coffee Cups—air pressure

An Imploding Can—air pressure

A Boiled Egg Sucked into a Bottle—atmospheric pressure

A Fantastic Egg Drop—inertia

An Egg That Stands—balance of forces, center of gravity

V. SHOW OFF YOUR ESP

A Swinging Pendulum—resonance

Buildings and Earthquakes—resonance

Snapping a Thread—inertia

Reading Words Through an Envelope—light reflection

VI. AMAZING SCIENCE TRICKS WITH THE BODY

The Iron Finger—balance of forces

The Shrinking Arm—functioning of muscles

Stretching Your Backbone—functioning of muscles

VII. AMAZING SCIENCE TRICKS IN THE LIVING ROOM

A Musical Straw—acoustic resonance

A Hanging Belt—balance of forces, center of gravity

A Skewered Balloon—polymer materials

A Floating Ping-pong Ball—air flow

How to Balance Anything—balance of forces, center of gravity

A Dancing Butterfly—static electricity

Lifting a Child with Your Breath—air pressure

Zero Gravity in a Paper Cup—weightlessness

Pinhole Glasses for Nearsightedness—linear motion of light

An On-again, Off-again Radio—electromagnetism

An Immovable Business Card—air flow

Hot Balloon, Cold Balloon—expansion and contraction of air